Connected  Mathematics™

Frogs, Fleas, and Painted Cubes

Quadratic Relationships

Student Edition

THE O. HENRY SCHOOL
I.S. 70 MAN.
333 west 17th street
New York, New York 10011

Glenda Lappan
James T. Fey
William M. Fitzgerald
Susan N. Friel
Elizabeth Difanis Phillips

Developed at Michigan State University

DALE SEYMOUR PUBLICATIONS®

Connected Mathematics™ was developed at Michigan State University with the support of National Science Foundation Grant No. MDR 9150217.

This project was supported, in part, by the
National Science Foundation
Opinions expressed are those of the authors and not necessarily those of the Foundation

The Michigan State University authors and administration have agreed that all MSU royalties arising from this publication will be devoted to purposes supported by the Department of Mathematics and the MSU Mathematics Education Enrichment Fund.

This book is published by Dale Seymour Publications®, an imprint of Addison Wesley Longman, Inc.

Managing Editor: Catherine Anderson
Project Editor: Stacey Miceli
Production/Manufacturing Director: Janet Yearian
Production/Manufacturing Coordinator: Claire Flaherty
Design Manager: John F. Kelly
Photo Editor: Roberta Spieckerman
Design: PCI, San Antonio, TX
Composition: London Road Design, Palo Alto, CA
Illustrations: Pauline Phung, Margaret Copeland, Ray Godfrey
Cover: Ray Godfrey

Photo Acknowledgements: 5 © The Bettman Archives; 7 © Gene Ahrens/FPG International; 21 © Mali Apple; 28 © Jean Morman Unsworth; 41 © Dean Abramson/Stock, Boston; 54 © Jeffrey Sylvester/FPG International; 55 © Corbis/Bettmann Archives; 56 © Blake Sell/Rueters/Bettmann; 61 © Bob Daemmrich/Stock, Boston; 65 © Michael Mauney/Tony Stone Images; 71 © Bryce Flynn/Stock, Boston

Copyright © 1998 by Michigan State University, Glenda Lappan, James T. Fey, William M. Fitzgerald, Susan N. Friel, and Elizabeth D. Phillips. All rights reserved. No part of this publication may be reproduced, stored in a retrieval system, or transmitted, in any form or by any means, electronic, mechanical, photocopying, recording, or otherwise, without prior written permission of the publisher. Printed in the United States of America.

Rubik's is a trademark of Seven Towns Ltd.

Order number 21479
ISBN 1-57232-184-9

5 6 7 8 9 10-BA-01 00 99

The Connected Mathematics Project Staff

Project Directors

James T. Fey
University of Maryland

William M. Fitzgerald
Michigan State University

Susan N. Friel
University of North Carolina at Chapel Hill

Glenda Lappan
Michigan State University

Elizabeth Difanis Phillips
Michigan State University

Project Manager

Kathy Burgis
Michigan State University

Technical Coordinator

Judith Martus Miller
Michigan State University

Curriculum Development Consultants

David Ben-Chaim
Weizmann Institute

Alex Friedlander
Weizmann Institute

Eleanor Geiger
University of Maryland

Jane Mitchell
University of North Carolina at Chapel Hill

Anthony D. Rickard
Alma College

Collaborating Teachers/Writers

Mary K. Bouck
Portland, Michigan

Jacqueline Stewart
Okemos, Michigan

Graduate Assistants

Scott J. Baldridge
Michigan State University

Angie S. Eshelman
Michigan State University

M. Faaiz Gierdien
Michigan State University

Jane M. Keiser
Indiana University

Angela S. Krebs
Michigan State University

James M. Larson
Michigan State University

Ronald Preston
Indiana University

Tat Ming Sze
Michigan State University

Sarah Theule-Lubienski
Michigan State University

Jeffrey J. Wanko
Michigan State University

Evaluation Team

Mark Hoover
Michigan State University

Diane V. Lambdin
Indiana University

Sandra K. Wilcox
Michigan State University

Judith S. Zawojewski
National-Louis University

Teacher/Assessment Team

Kathy Booth
Waverly, Michigan

Anita Clark
Marshall, Michigan

Julie Faulkner
Traverse City, Michigan

Theodore Gardella
Bloomfield Hills, Michigan

Yvonne Grant
Portland, Michigan

Linda R. Lobue
Vista, California

Suzanne McGrath
Chula Vista, California

Nancy McIntyre
Troy, Michigan

Mary Beth Schmitt
Traverse City, Michigan

Linda Walker
Tallahassee, Florida

Software Developer

Richard Burgis
East Lansing, Michigan

Development Center Directors

Nicholas Branca
San Diego State University

Dianne Briars
Pittsburgh Public Schools

Frances R. Curcio
New York University

Perry Lanier
Michigan State University

J. Michael Shaughnessy
Portland State University

Charles Vonder Embse
Central Michigan University

Special thanks to the students and teachers at these pilot schools!

Baker Demonstration School
Evanston, Illinois

Bertha Vos Elementary School
Traverse City, Michigan

Blair Elementary School
Traverse City, Michigan

Bloomfield Hills Middle School
Bloomfield Hills, Michigan

Brownell Elementary School
Flint, Michigan

Catlin Gabel School
Portland, Oregon

Cherry Knoll Elementary School
Traverse City, Michigan

Cobb Middle School
Tallahassee, Florida

Courtade Elementary School
Traverse City, Michigan

Duke School for Children
Durham, North Carolina

DeVeaux Junior High School
Toledo, Ohio

East Junior High School
Traverse City, Michigan

Eastern Elementary School
Traverse City, Michigan

Eastlake Elementary School
Chula Vista, California

Eastwood Elementary School
Sturgis, Michigan

Elizabeth City Middle School
Elizabeth City, North Carolina

Franklinton Elementary School
Franklinton, North Carolina

Frick International Studies Academy
Pittsburgh, Pennsylvania

Gundry Elementary School
Flint, Michigan

Hawkins Elementary School
Toledo, Ohio

Hilltop Middle School
Chula Vista, California

Holmes Middle School
Flint, Michigan

Interlochen Elementary School
Traverse City, Michigan

Los Altos Elementary School
San Diego, California

Louis Armstrong Middle School
East Elmhurst, New York

McTigue Junior High School
Toledo, Ohio

National City Middle School
National City, California

Norris Elementary School
Traverse City, Michigan

Northeast Middle School
Minneapolis, Minnesota

Oak Park Elementary School
Traverse City, Michigan

Old Mission Elementary School
Traverse City, Michigan

Old Orchard Elementary School
Toledo, Ohio

Portland Middle School
Portland, Michigan

Reizenstein Middle School
Pittsburgh, Pennsylvania

Sabin Elementary School
Traverse City, Michigan

Shepherd Middle School
Shepherd, Michigan

Sturgis Middle School
Sturgis, Michigan

Terrell Lane Middle School
Louisburg, North Carolina

Tierra del Sol Middle School
Lakeside, California

Traverse Heights Elementary School
Traverse City, Michigan

University Preparatory Academy
Seattle, Washington

Washington Middle School
Vista, California

Waverly East Intermediate School
Lansing, Michigan

Waverly Middle School
Lansing, Michigan

West Junior High School
Traverse City, Michigan

Willow Hill Elementary School
Traverse City, Michigan

Contents

Mathematical Highlights	4
Investigation 1: Introduction to Quadratic Relationships	5
1.1 Staking a Claim	6
1.2 Reading a Graph	7
1.3 Writing an Equation	10
Applications—Connections—Extensions	12
Mathematical Reflections	18
Investigation 2: Quadratic Expressions	19
2.1 Trading Land	20
2.2 Changing One Dimension	22
2.3 Changing Both Dimensions	24
2.4 Looking Back at Parabolas	28
Applications—Connections—Extensions	31
Mathematical Reflections	40
Investigation 3: Quadratic Patterns of Change	41
3.1 Counting Handshakes	41
3.2 Exploring Triangular Numbers	43
Applications—Connections—Extensions	45
Mathematical Reflections	51
Investigation 4: What Is a Quadratic Function?	52
4.1 Tracking a Ball	53
4.2 Measuring Jumps	55
4.3 Putting It All Together	57
Applications—Connections—Extensions	60
Mathematical Reflections	70
Investigation 5: Painted Cubes	71
5.1 Analyzing Cube Patterns	71
5.2 Exploring Painted-Cube Patterns	73
Applications—Connections—Extensions	75
Mathematical Reflections	84
Glossary	85
Index	88

Frogs, Fleas, and Painted Cubes

A ball is thrown into the air. The height of the ball in feet after t seconds can be described by the equation $h = -16t^2 + 64t$. What is the maximum height reached by the ball? When does the ball reach this maximum height?

Imagine that you are prospecting diamonds in the barren outback of western Australia. You are told that you can claim any rectangular piece of land that can be surrounded by 60 meters of fencing. How should you arrange your fencing in order to enclose the maximum area possible?

After a victory, members of a winning team may congratulate each other with a round of high fives. How many high fives would be exchanged among a team with 5 players? Among a team with 6 players? Among a team with n players?

Mathematics is useful for solving practical problems in science, business, engineering, and economics. In earlier units, you studied problems that could be modeled with linear or exponential relationships. In this unit, you will explore quadratic relationships. Quadratic relationships are found in many interesting situations including those on the opposite page.

Mathematical Highlights

In this unit, you will explore quadratic relationships.

- As you find dimensions of rectangles with a perimeter of 20 meters, you are introduced to quadratic relationships.

- Examining land trades helps you learn to write quadratic expressions in both expanded form and factored form.

- As you explore interesting quadratic relationships, you learn about the important features of *parabolas,* the graphs of quadratic relationships.

- Comparing graphs and equations for quadratic relationships helps you see that many features of a graph can be predicted from its equation.

- As you solve counting problems involving handshakes, high fives, and dot patterns, you explore the patterns of change for quadratic functions.

- Exploring how the height of a ball changes when it is thrown and how the heights of a frog, a flea, and a basketball player change when they jump gives you practice interpreting graphs, tables, and equations for quadratic functions.

- Computing first and second differences for tables of quadratic functions helps you better understand the patterns of change for quadratic relationships.

- Investigating the patterns involved in cube puzzles gives you practice recognizing linear and quadratic functions.

Using a Calculator

In this unit, you will use a graphing calculator to graph quadratic equations. As you work on the Connected Mathematics™ units, you decide whether to use a calculator to help you solve a problem.

INVESTIGATION 1

Introduction to Quadratic Relationships

When Mexico ceded California to the United States in 1848, California was a relatively unexplored territory with only a few thousand people. Then, in January of 1848, gold was discovered at Sutter's mill near Sacramento. By 1849, a great gold rush had begun. The gold rush brought 500,000 new residents to California. By 1850, California had enough people to be admitted to the Union as a state. Today, California has the greatest population of any U.S. state.

Throughout history, people have traveled to particular areas of the world with hopes of getting rich.

- In 1871, prospectors headed to South Africa in search of diamonds.
- From 1860 to 1900, farmers moved to the American prairie where land was free to homesteaders.
- The 1901 Spindletop oil gusher brought drillers by the thousands to eastern Texas.

The prospectors and farmers had to stake claims on the land they wanted to work. Those who made smart claims knew the land . . . and some mathematics. Surveying, determining acreage, buying fencing, and analyzing costs all required the tools of mathematics.

1.1 Staking a Claim

Imagine that it is the year 2100 and a rare and precious metal has just been discovered on the planet Mars. You and hundreds of other adventurers are traveling to the planet to stake your claim. Each new prospector is allowed to claim any piece of land that can be surrounded by 20 meters of laser fencing. You want to arrange your fencing to enclose the maximum area possible.

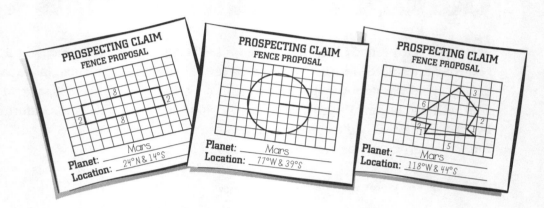

Problem 1.1

Suppose the Mars colony adds the restriction that each claim must be rectangular.

A. Sketch several rectangles with a perimeter of 20 meters. Include some with small areas and some with large areas. Label the dimensions of each rectangle.

B. Make a table showing the length of a side and the area for each rectangle with a perimeter of 20 meters and whole-number side lengths.

C. Make a graph of your (length of a side, area) data. Describe the shape of the graph.

D. If you want to enclose the greatest area possible with your fencing, what should the dimensions of your fence be? How can you use your graph to justify your answer?

Problem 1.1 Follow-Up

1. Suppose the dimensions of the rectangle were not restricted to whole numbers. Would this change your answer to part D? Explain.

2. Suppose the shape of a claim were not restricted to a rectangle. How could you arrange your 20 meters of fencing to enclose a greater area?

Frogs, Fleas, and Painted Cubes

1.2 Reading a Graph

The relationship between length and area in Problem 1.1 is an example of a **quadratic relationship**. Quadratic relationships are characterized by their U-shaped graphs, which are called **parabolas**. In this relationship, the area depends on, or is a *function* of, the length. We refer to a relationship in which one variable depends on another as a **function**.

Many of the linear and exponential relationships you studied in earlier units are functions. For example, if a van travels at a steady rate, the distance it covers is a function of the travel time. The relationship between travel time and distance is a linear function. The value of an investment that grows by 4% per year is a function of the number of years. The relationship between the number of years and the value is an exponential function.

As you explore quadratic functions in this unit, look for common patterns in tables, graphs, and equations.

The Gateway Arch in St. Louis, Missouri, resembles a parabola.

Investigation 1: Introduction to Quadratic Relationships

Problem 1.2

The graph below shows length and area data for rectangles with a fixed perimeter.

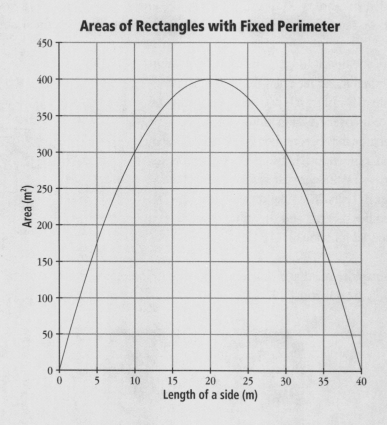

A. Describe the shape of the graph and any special features you observe.

B. What is the greatest area possible for a rectangle with this perimeter? What are the dimensions of this rectangle?

C. What is the area of the rectangle with a side of length 12 meters? What is the area of the rectangle with a side of length 28 meters? Explain how these two rectangles are related.

D. What are the dimensions of the rectangle with an area of 300 square meters?

E. What is the fixed perimeter for the rectangles represented by the graph? Explain how you found the perimeter.

Problem 1.2 Follow-Up

The table below shows length and area data for rectangles with a fixed perimeter.

Length of a side (m)	Area (m²)
0	0
1	11
2	20
3	27
4	32
5	35
6	36
7	35
8	32
9	27
10	20
11	11
12	0

1. How do the shape and the special features you observed for the graph in Problem 1.2 appear in the table?

2. What is the fixed perimeter for the rectangles represented in this table? Explain how you found the perimeter.

3. What is the greatest area possible for a rectangle with this perimeter? What are the dimensions of this rectangle?

4. Approximate the dimensions of a rectangle with this fixed perimeter and an area of 16 square meters.

1.3 Writing an Equation

In Problems 1.1 and 1.2, you looked at tables and graphs of relationships between length and area for rectangles with fixed perimeters. In this problem, you will write equations for these relationships.

Problem 1.3

The rectangle below has a perimeter of 20 meters and a side of length l meters.

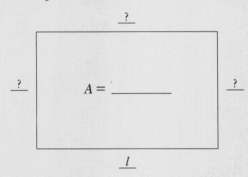

A. Express the length of each side of the rectangle in terms of l. That is, write an expression that contains the variable l to represent the length of each side.

B. Write an equation for the area, A, of the rectangle in terms of l.

C. If the length of a side of the rectangle is 6 meters, what is the area?

D. Use a calculator to make a table and a graph for your equation. Show x values from 1 to 10 and y values from 0 to 30. Compare your table and graph to those you made in Problem 1.1.

■ Problem 1.3 Follow-Up

1. Consider rectangles with a perimeter of 60 meters.
   a. As in Problem 1.3, draw a rectangle to represent this situation. Label one side l, and label the other sides in terms of l.
   b. Write an equation for the area, A, in terms of l.
   c. Make a table for your equation. Then, use your table to estimate the greatest area possible for a rectangle with a perimeter of 60 meters. Give the side lengths of this rectangle.
   d. Use a calculator or data from your table to help you sketch a graph of the relationship between the length of a side and the area.
   e. How can you use your graph to find the maximum area? How does your graph show the side length that corresponds to the maximum area?

2. An equation for the area of rectangles with a certain fixed perimeter is $A = l(35 - l)$.
   a. Draw a rectangle to represent this situation. Label one side l, and label the other sides in terms of l.
   b. If the length of a side of a rectangle with this fixed perimeter is 20 meters, what is the area?
   c. Describe two ways you could find the perimeter for the rectangles represented by this equation. What is the perimeter?
   d. Describe the graph of this equation.
   e. What is the maximum area for this family of rectangles? What dimensions correspond to this maximum area? Explain how you found your answers.

3. If you know the perimeter of a rectangle, how can you write an equation for the area in terms of the length of a side?

4. Graphs of quadratic functions are called *parabolas*. Describe the characteristics of the parabolas you have seen so far.

5. Study the graphs, tables, and equations for areas of rectangles with fixed perimeters. Which representation is most useful for predicting the maximum area?

Applications • Connections • Extensions

As you work on these ACE questions, use your calculator whenever you need it.

Applications

1. This rectangle has a perimeter of 30 meters and a side of length l meters.

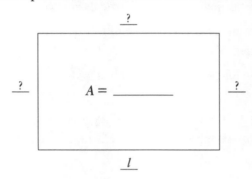

a. Express the lengths of the other sides in terms of l.

b. Write an equation for the area, A, in terms of l.

c. Make a graph of your equation, and describe its shape.

d.   **i.** Use your equation to find the area of the rectangle if the length of one side is 10 meters.

    ii. Describe how you could use your graph to find the area of the rectangle if the length of one side is 10 meters.

    iii. Describe how you could use a table to find the area of the rectangle if the length of one side is 10 meters.

e. What is the maximum area possible for a rectangle with a perimeter of 30 meters? What are the dimensions of the rectangle with the maximum area?

2. This rectangle has a perimeter of 50 meters and a side of length *l* meters.

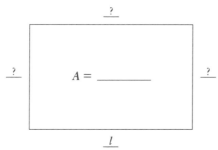

a. Express the lengths of the other sides in terms of *l*.

b. Write an equation for the area, *A*, in terms of *l*.

c. Make a graph of your equation, and describe its shape.

d. **i.** Use your equation to find the area of the rectangle if the length of one side is 10 meters.

  ii. Describe how you could use your graph to find the area of the rectangle if the length of one side is 10 meters.

  iii. Describe how you could use a table to find the area of the rectangle if the length of one side is 10 meters.

e. What is the maximum area for a rectangle with a perimeter of 50 meters? What are the dimensions of this rectangle?

3. Find the maximum area for a rectangle with a perimeter of 120 meters. Make your answer convincing by including these things:

- Sketches of rectangles with a perimeter of 120 meters that do not have the maximum area and a sketch of the rectangle you think does have the maximum area

- A table of lengths and areas for rectangles with a perimeter of 120 meters, using increments of 5 meters for the lengths

- A graph of the relationship between length and area

Explain how each piece of evidence supports your answer.

Investigation 1: Introduction to Quadratic Relationships

4. What is the maximum area for a rectangle with a perimeter of 130 meters? As in ACE question 3, support your answer with sketches of rectangles, a table, and a graph. In your table, use increments of 2.5 for the lengths.

5. The graph below shows lengths and areas for rectangles with a fixed perimeter.

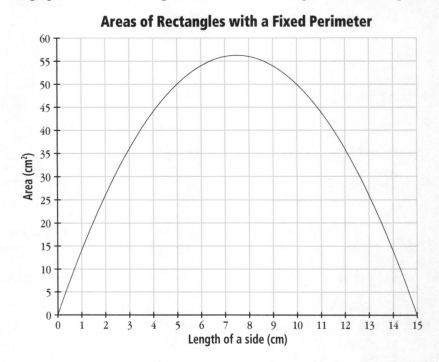

a. Describe the shape of the graph and any special features you observe.

b. What is the maximum area for a rectangle with this fixed perimeter? What are the dimensions of this rectangle?

c. As the length of a side gets very close to 15 centimeters, what happens to the area?

d. What is the area of a rectangle with this fixed perimeter and a side of length 3 centimeters?

e. Describe two ways you could find the fixed perimeter for the rectangles represented by the graph.

6. An equation for the area in square meters of rectangles with a fixed perimeter is $A = l(25 - l)$.

   a. Describe the graph of this equation.

   b. What is the maximum area for a rectangle with this perimeter? What dimensions correspond to this area? Explain how you found your answers.

   c. What is the area of a rectangle with this fixed perimeter and a side of length 20 meters?

   d. Describe two ways to find the fixed perimeter for rectangles represented by this equation.

7. This incomplete graph shows data for rectangles with a fixed perimeter.

Areas of Rectangles with a Fixed Perimeter

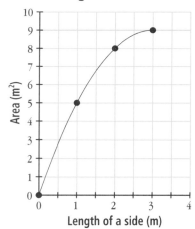

   a. Copy and complete the graph to show areas for rectangles with side lengths greater than 3 meters. Explain why your graph is correct.

   b. Make a table of data for this situation.

   c. What is the maximum area for a rectangle with this perimeter? What are the dimensions of this rectangle? Explain how you found your answers.

   d. Which of these equations describes the graph?

   $A = l(l - 6)$ $A = l(12 - l)$ $A = l(6 - l)$ $A = l(3 - l)$

Investigation 1: Introduction to Quadratic Relationships

8. This table contains data for rectangles with a fixed perimeter.

Length of a side (m)	Area (m²)
0	0
1	7
2	12
3	15
4	16
5	
6	
7	
8	

a. Copy and complete the table. Explain why you think your table is correct.

b. Make a graph of the relationship between length and area for rectangles with this perimeter.

c. What are the dimensions of the rectangle with maximum area?

d. Which of these equations describes the graph?

$A = l(8 - l)$ $A = l(16 - l)$ $A = l(4 - l)$ $A = l(l - 8)$

Connections

9. a. Suppose a rectangular field has a perimeter of 300 yards. The relationship between the length and the width of the field can be represented by the equation $l = 150 - w$. Is the relationship between the length and the width quadratic? How do you know?

b. Suppose a field is a nonrectangular parallelogram with a perimeter of 300 yards. Is the relationship between the side lengths the same as it is for the rectangular field in part a? How do you know?

c. Suppose a field is a quadrilateral that is not a parallelogram. The perimeter of the field is 300 yards. Is the relationship between the side lengths the same as it is for the rectangular field in part a? How do you know?

10. Ms. Carillo, a photographer, sells prints of her photographs. She would like to increase her profit by raising the price of the prints. However, she knows if the price is too high, she won't sell as many prints and her profit will decrease. Assume the following equation gives the monthly profit, P, she would earn if she charges d dollars for each print:

$$P = d(100 - d)$$

   a. Make a table and a graph for this equation. Use increments of 10 dollars for the price.

   b. Estimate the price that will produce the maximum profit. Explain how you found your answer.

   c. How are the table and the graph for this situation similar to the table and graph you made in Problem 1.3? How are they different?

   d. How is the equation similar to and different from the equation you wrote in Problem 1.3?

Extensions

11. Of all rectangles with whole-number side lengths and an area of 20 square centimeters, which has the smallest perimeter? Give evidence for your answer.

12. Mr. DeAngelo is designing a school building. The floor of the music room will be a rectangle with an area of 1200 square feet.

   a. Make a table with about ten rows showing a range of possible lengths and widths for the music room floor.

   b. Add a column to your table for the perimeter of each rectangle.

   c. What pattern do you see in the perimeter column? What kinds of rectangles have large perimeters and what kinds of rectangles have small perimeters?

   d. Write an equation you could use to calculate the length of the floor for any given width.

Mathematical Reflections

In this investigation, you found the maximum area for rectangles with a fixed perimeter. The relationship between length and area for rectangles with a fixed perimeter is an example of a quadratic function. These questions will help you summarize what you have learned:

1. In a function relating two variables, one variable *depends* on the other variable. For a function relating length and area for rectangles with a fixed perimeter, which variable is the dependent variable? Which variable is the independent variable?

2. The graph of a quadratic function is a parabola.

    a. What patterns did you observe in the parabolas in this investigation?

    b. How do the patterns in a parabola appear in the table of values for a quadratic function?

3. Describe two ways to find the maximum area for rectangles with a fixed perimeter.

4. How are tables, graphs, and equations for quadratic functions different from those for linear and exponential functions?

Think about your answers to these questions, discuss your ideas with other students and your teacher, and then write a summary of your findings in your journal.

INVESTIGATION 2

Quadratic Expressions

If you gave a friend two $1 bills, and your friend gave you eight quarters, you would consider it a fair trade. If you left your job an hour early one day and worked an hour later the next day, your boss would probably consider it a fair trade. Sometimes it is not as easy to determine whether a trade is fair.

Think about this!

U.S. Malls Incorporated wants to build a new shopping center. The mall developer has bought all the land on the proposed site except for one square lot that measures 125 meters on each side. The family that owns the lot is reluctant to sell it. In exchange for the lot, the developer has offered to give the family a rectangular lot that is 100 meters longer on one side and 100 meters shorter on another side than the square lot. Do you think this is a fair trade?

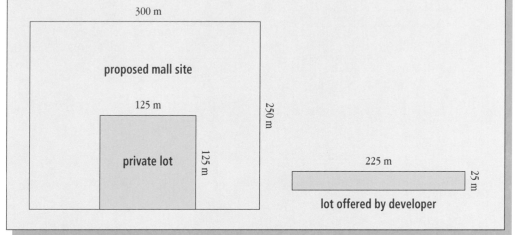

2.1 Trading Land

In this problem, you will explore whether a trade like the one offered by U.S. Malls Incorporated will always be fair, sometimes be fair, or never be fair. You will look at a simple trade situation to see if you can discover a pattern that will help you make predictions about more complex situations.

> **Problem 2.1**
>
> Suppose you own a square piece of land with sides n meters long. You trade your land for a rectangular lot. The length of your new lot is 2 meters longer than the side length of your original lot, and the width of your new lot is 2 meters shorter than the side length of the original lot.
>
>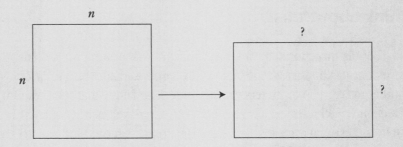
>
> **A.** Copy and complete the table below.
>
> | Original square | | New rectangle | | | Difference |
> |---|---|---|---|---|---|
> | Side length (m) | Area (m²) | Length (m) | Width (m) | Area (m²) | in areas (m²) |
> | 3 | 9 | 5 | 1 | 5 | 4 |
> | 4 | | | | | |
> | 5 | | | | | |
> | 6 | | | | | |
> | 7 | | | | | |
>
> **B.** For each side length in the table, tell how the area of your new lot compares with the area of the original lot. For which side lengths, if any, is this a fair trade?
>
> **C.** The side length of the original square lot was n meters. For each column in the table, write an expression for the values in the column in terms of n. For example, the expression for the area of the original square is n^2.

20 Frogs, Fleas, and Painted Cubes

Problem 2.1 Follow-Up

Symbolic rules such as n^2 are often called *symbolic expressions,* or simply *expressions.* When you use symbols to show what an expression represents, you get an *equation.* If we let A be the area of the original lot, we can write the equation $A = n^2$.

1. a. Graph the equation for the relationship between the side length and the area of the original lot. Use n values from $^-10$ to 10 and A values from $^-20$ to 100.

b. Write an equation for the relationship between the side length of the original lot and the area of the new lot. Graph the equation using the same window settings as before.

c. Describe the shapes of your graphs.

d. Do all the values for area and side length on your graph make sense? Why or why not?

2. a. Compare the table of side lengths and areas for the lots of land with the tables of lengths and areas for rectangles with fixed perimeters from Investigation 1. How are the tables similar? How are they different?

b. Compare the graphs of side lengths and areas for the lots of land with the graphs of lengths and areas for rectangles with fixed perimeters from Investigation 1. How are the graphs similar? How are they different?

c. Compare the equations for the areas of the lots of land with the equations for the areas of rectangles with fixed perimeters from Investigation 1. How are the equations similar? How are they different?

3. Look back at the situation discussed in the "Think about this!" box on page 19. Is the trade offered by U.S. Malls Incorporated a good idea for the family that owns the private lot? Explain your answer.

Investigation 2: Quadratic Expressions 21

2.2 Changing One Dimension

In mathematics, there is often more than one way to express a quantity. For example, 2 + 5, 3 + 4, and 7 are equivalent expressions. In this problem, you will use diagrams to help you write equivalent expressions for the area of a rectangle.

Problem 2.2

A. A square has sides of length x centimeters. A new rectangle is created by increasing one dimension of the square by 2 centimeters.

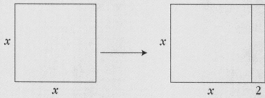

   1. The new rectangle is made up of the original square and an added rectangle. What are the dimensions of the added rectangle? What is its area?

   2. Write an equation for the area of the new rectangle as the sum of the area of the original square and the area of the added rectangle.

   3. What are the length and the width of the new rectangle? Write an equation for the area of the new rectangle as its length times its width.

   4. Graph your equations from parts 2 and 3 on your calculator, and copy the graphs onto your paper. Describe the shapes of the graphs. How do the graphs compare? What does this tell you about the two equations?

B. A square has sides of length x centimeters. One dimension of the square is increased by 3 centimeters to create a new rectangle.

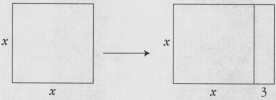

   1. How much greater is the area of the new rectangle than the area of the square?

   2. Write two equations for the area of the new rectangle.

   3. Graph both equations on your calculator, and copy the graphs onto your paper. Describe the shapes of the graphs. How do the graphs compare?

Frogs, Fleas, and Painted Cubes

■ **Problem 2.2 Follow-Up**

You can express the area of the rectangular lot in Problem 2.1 in two ways. You can write $(n - 2)(n + 2)$ to express the area as the product of the length and the width, or you can write $n^2 - 4$ to express the fact that the area of the rectangular lot is 4 square meters less than the area of the square lot.

The expression $(n - 2)(n + 2)$ is said to be in **factored form** because it is written as the product of two linear factors. An expression that is written as the sum or difference of terms is said to be in *term form,* or **expanded form**. A **term** is an expression that consists of variables and/or numbers multiplied together. The expression $n^2 - 4$ is in expanded form because it is written as the difference of the terms n^2 and 4. Since $(n - 2)(n + 2)$ and $n^2 - 4$ are equivalent, you can write $(n - 2)(n + 2) = n^2 - 4$.

In part A of Problem 2.2, you expressed the area of a rectangle in two equivalent ways. The factored form, $x(x + 2)$, is the product of the two linear factors x and $x + 2$. The equivalent expanded form is $x^2 + 2x$.

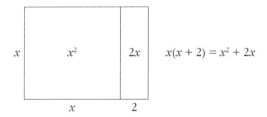

1. The diagram below shows a rectangle divided into two smaller rectangles. Write two expressions, one in factored form and one in expanded form, for the area of the large rectangle.

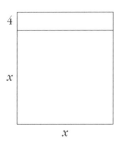

Investigation 2: Quadratic Expressions 23

2. Write two expressions, one in factored form and one in expanded form, for the area of the unshaded region.

3. Each expression below represents the area of a rectangle. Do parts a and b for each expression.

$x(x + 4)$ $x(x - 4)$ $x(5 + 2)$

  a. Draw a rectangle divided to show that its area is represented by the expression. Label the lengths and areas on your drawing.
  b. Write an equivalent expression in expanded form.

4. Each expression below represents the area of a rectangle. Do parts a and b for each expression.

$x^2 + 5x$ $x^2 - 5x$ $5x + 4x$

  a. Draw a rectangle divided to show that its area is represented by the expression. Label the lengths and areas on your drawing.
  b. Write an equivalent expression in factored form.

5. An equation that describes a quadratic function is called a *quadratic equation*. If one side of an equation is not written in factored form, how can you tell whether it is a quadratic equation?

2.3 Changing Both Dimensions

In Problem 2.2, you looked at how the area changes when one dimension of a square is increased or decreased to form a new rectangle. You will now explore what happens to the area when *both* dimensions of a square are increased to form a new rectangle.

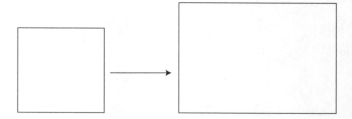

Problem 2.3

A. A square has sides of length *x* centimeters. A new rectangle is created by increasing one dimension of the square by 2 centimeters and increasing the other dimension by 3 centimeters.

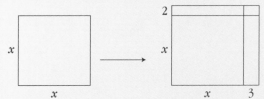

1. Copy the new rectangle. Label the area of each of the four sections.

2. Write two expressions, one in factored form and one in expanded form, for the area of the new rectangle.

3. Use your expressions from part 2 to write two equations for the area, *A*, of the rectangle. Graph both equations on your calculator. Compare these graphs with the graphs you made in Problem 2.2.

B. A square has sides of length *x* centimeters. One dimension of the square is doubled and then increased by 2 centimeters, and the other dimension is increased by 3 centimeters.

1. Make a sketch to show how the square is transformed into the new rectangle. Label the area of each section of the new rectangle.

2. Write two expressions, one in factored form and one in expanded form, for the area of the new rectangle.

3. Use your expressions from part 2 to write two equations for the area, *A*, of the rectangle. Graph both equations on your calculator. Compare these graphs with the graphs you made in Problem 2.2.

C. The rectangle below is divided into four smaller rectangles. Write two expressions, one in factored form and one in expanded form, for the area of the large rectangle.

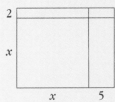

Investigation 2: Quadratic Expressions 25

Problem 2.3 Follow-Up

1. In a–c, write two expressions, one in factored form and one in expanded form, for the area of the entire figure.

   a.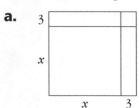

   b.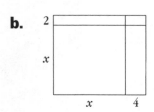

   c.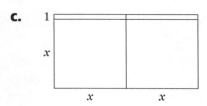

2. One dimension of a square is increased by 1 centimeter, and the other dimension is increased by 3 centimeters. Write two expressions, one in factored form and one in expanded form, for the area of the new rectangle.

3. One dimension of a square is doubled, and the other dimension is increased by 4 centimeters. Write two expressions, one in factored form and one in expanded form, for the area of the new rectangle.

4. Do parts a and b for each expression.

   $(x + 3)(x + 4)$ $(x + 5)(x + 5)$ $2x(x + 2)$

   a. Draw and label a rectangle whose area is represented by the expression.
   b. Write an equivalent expression in expanded form.

The area of the rectangle below can be written in expanded form as $x^2 + 2x + 3x + 6$. Since $2x$ and $3x$ have the same variable with the same exponent (in this case the exponent is 1), they are called **like terms.** You can combine like terms to get a single term. In the expression $x^2 + 2x + 3x + 6$, you can combine $2x$ and $3x$ to get $5x$. The expression then becomes $x^2 + 5x + 6$.

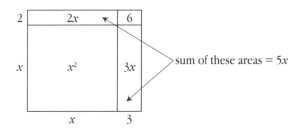

Sometimes it is helpful to combine like terms; other times it is helpful to break a term into two terms. For example, you can write $8x$ as $7x + x$, $6x + 2x$, $5x + 3x$, and so on.

5. a. Copy the diagram below, replacing each question mark with the correct length or area.

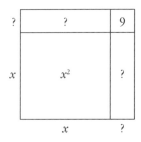

   b. Write two expressions for the area of the large rectangle.

6. a. Draw a rectangle made up of four smaller rectangles similar to the one in question 5. Make the area of the square in the lower left corner x^2, and make the area of the rectangle in the upper right corner 8. What might the areas of the other two small rectangles be? Find one solution, and use it to label the lengths and areas in your diagram.

   b. Write two expressions for the area of the large rectangle.

7. Do parts a and b for each expression below.

   $x^2 + x + 5x + 5$ $x^2 + 7x + 6$ $x^2 + 8x + 16$

   a. Draw and label a rectangle whose area is represented by the expression.
   b. Write an equivalent expression in factored form.

Investigation 2: Quadratic Expressions

2.4 Looking Back at Parabolas

All the functions you have studied in Investigations 1 and 2 are quadratic functions. The graphs of these functions are parabolas. The parabolas in Investigation 1 are upside-down U-shapes, and the parabolas in Investigation 2 are right-side-up U-shapes.

If you draw a vertical line through the maximum point or minimum point of a parabola and then fold along this line, the two halves of the parabola will exactly match. This vertical line is called the **line of symmetry** for the graph.

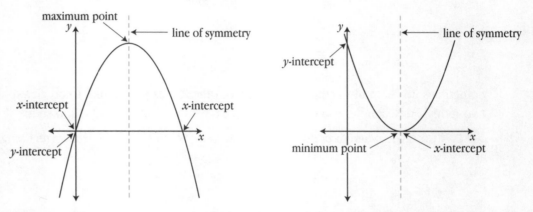

Many important patterns occur in graphs, tables, and equations of quadratic functions. In this problem, you will explore these questions:

What can you learn about a quadratic function from its graph?

How are the features of a parabola related to its equation?

The St. Louis Priori Chapel has windows shaped like parabolas.

Frogs, Fleas, and Painted Cubes

Problem 2.4

The eight equations below were graphed on a calculator using the window settings shown. The graphs shown below and on the next page are reproduced on Labsheet 2.4.

$y = x^2$ $y = x(x + 4)$

$y = x(4 - x)$ $y = (x + 3)(x - 3)$

$y = (x + 3)(x + 3)$ $y = (x + 2)(x + 3)$

$y = x(x - 4)$ $y = 2x(x + 4)$

```
WINDOW
XMIN=-5
XMAX=5
XSCL=1
YMIN=-10
YMAX=10
YSCL=1
```

Do parts A–F for each equation.

A. Match the equation to its graph.

B. Label the coordinates of the *x*-intercepts on the graph. Describe how you can predict the *x*-intercepts from the equation.

C. Draw the line of symmetry on the graph.

D. Describe the shape of the graph, and label the coordinates of the maximum or minimum point.

E. Tell what features of the graph you can predict from the equation.

F. Draw and label a rectangle whose area is represented by the equation. Then, express the area of the rectangle in expanded form.

Graph 1

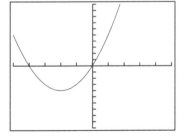

Graph 2

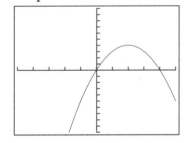

Graph 3

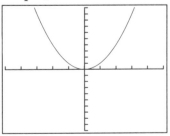

Graph 4

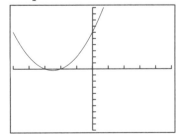

Graph 5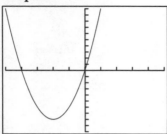

Graph 6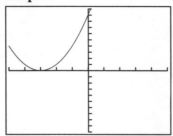

Graph 7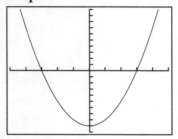

Graph 8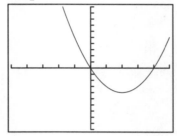

■ **Problem 2.4 Follow-Up**

1. If an equation is written in factored form, how can you tell whether it represents a quadratic function?

2. If an equation is written in expanded form, how can you tell whether it represents a quadratic function?

Applications • Connections • Extensions

As you work on these ACE questions, use your calculator whenever you need it.

Applications

1. A square has sides of length x centimeters. A new rectangle is created by increasing one dimension by 4 centimeters and decreasing the other dimension by 4 centimeters.

  a. Make a table showing the area of the square and the area of the new rectangle for whole-number x values from 4 to 16.

  b. On the same set of axes, graph the (x, area) data for both the square and the rectangle.

  c. For which representation—the table or the graph—is it easier to compare the area of a square with the area of the new rectangle?

  d. Write an equation for the area of the square and an equation for the area of the new rectangle.

  e. Use your calculator to graph both equations for x values from $^-10$ to 10. Copy the graphs onto your paper.

2. A square has sides of length x centimeters. A new rectangle is created by increasing one dimension by 5 centimeters.

  a. Make a sketch to show how the square is transformed into the new rectangle.

  b. Write two expressions, one in factored form and one in expanded form, for the area of the new rectangle.

  c. Write an equation for the area, A, of the rectangle, and graph the equation.

Investigation 2: Quadratic Expressions

3. A square has sides of length x centimeters. A new square is created by increasing both dimensions by 5 centimeters.

   a. Make a sketch to show how the original square is transformed into the new square.

   b. Write two expressions, one in factored form and one in expanded form, for the area of the new square.

   c. Write an equation for the area, A, of the new square, and graph the equation.

4. A square has sides of length x centimeters. A new rectangle is created by increasing one dimension by 4 centimeters and increasing the other dimension by 5 centimeters.

   a. Make a sketch to show how the square is transformed into the new rectangle.

   b. Write two expressions, one in factored form and one in expanded form, for the area of the new rectangle.

   c. Write an equation for the area, A, of the new rectangle, and graph the equation.

In 5–9, write two expressions, one in factored form and one in expanded form, for the area of the unshaded region.

5.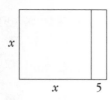

6.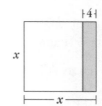

7.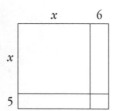

8.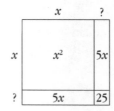

9.

In 10–14, the expression represents the area of a rectangle made by changing the dimensions of a square with sides of length x centimeters. Match the expression with the correct instructions.

	Area		Instructions for changing a square into a rectangle
10.	$(x-3)(x+3)$	**a.**	Increase one dimension by 3 centimeters, and increase the other dimension by 5 centimeters.
11.	$x(x+5)$	**b.**	Increase one dimension by 3 centimeters, and decrease the other dimension by 3 centimeters.
12.	$(x+3)(x+5)$	**c.**	Decrease one dimension by 5 centimeters, and increase the other dimension by 3 centimeters.
13.	$(x-3)(x+5)$	**d.**	Increase one dimension by 5 centimeters, and do not change the other dimension.
14.	$(x+3)(x-5)$	**e.**	Increase one dimension by 5 centimeters, and decrease the other dimension by 3 centimeters.

15. For each expression in ACE questions 10–14, write an equivalent expression in expanded form.

16. In ACE questions 10–14, which of the areas is always greater than the area of the original square? Explain how you found your answers.

17. For each expression in ACE questions 10–14, tell where the line of symmetry for the corresponding graph crosses the x-axis. For example, for question 10, tell where the line of symmetry for the graph of $A = (x-3)(x+3)$ crosses the x-axis.

Investigation 2: Quadratic Expressions

In 18–21, the equation represents the area of a rectangle made by changing the dimensions of a square with side lengths of x centimeters. Answer parts a–c.

18. $A = (x + 5)(x + 5)$

19. $A = x(x + 3)$

20. $A = (x + 1)(x + 3)$

21. $A = x(2x + 2)$

a. Make a sketch to show how the rectangle was created from the square.

b. Write an expression for the area in expanded form.

c. What are the x-intercepts of the graph of the equation? How can you determine the x-intercepts from the graph? How can you determine the x-intercepts from the equation?

In 22–26, based on what you have learned so far, tell whether the equation represents a quadratic relationship, a linear relationship, or an exponential relationship. Explain your answer.

22. $y = 5x + x^2$

23. $y = 2x + 8$

24. $y = (9 - x)x$

25. $y = 4x(x + 3)$

26. $y = 3^x$

27. Oletha made a rectangle from a square by doubling one dimension and then adding 3. She left the other dimension unchanged.

a. Write an equation for the area, A, of the new rectangle in terms of the side length, x, of the original square.

b. Graph your area equation. What are the x-intercepts of the graph? How can you find the x-intercepts from the graph? How can you find them from the equation?

Frogs, Fleas, and Painted Cubes

In 28–31, do parts a–c.

28. $y = (x + 7)(x + 2)$

29. $y = x(x + 3)$

30. $y = (x - 4)(x + 6)$

31. $y = (x - 5)(x + 5)$

a. Match the equation with its graph.

i.
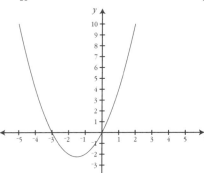

ii.
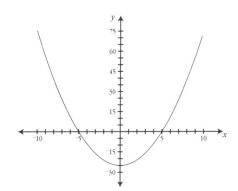

iii.
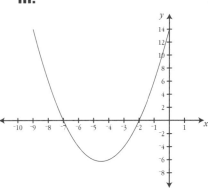

iv.
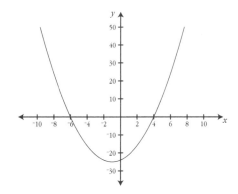

b. Tell where the line of symmetry for the graph crosses the *x*-axis.

c. Draw and label the rectangle whose area is represented by the equation.

Connections

32. The Stellar Cellular long-distance company charges $13.95 per month plus $0.39 per minute of calling time. The Call Anytime company has no monthly service fee but charges $0.95 per minute of calling time.

   a. When deciding which company to use, what factors should a potential customer consider?

   b. Describe each charge plan with an equation, a table, and a graph.

   c. For each plan, tell whether the relationship between calling time and monthly cost is quadratic. How do your equation, table, and graph support your answer?

33. The winner of the Jammin' Jelly jingle contest will receive $500. Antonia and her friends are writing a jingle for the contest. They plan to divide the prize money equally if they win.

   a. If n friends write the winning jingle, how much prize money will each person receive?

   b. Describe the relationship between the number of friends and the amount of prize money each friend receives.

   c. What kinds of questions about this relationship would be easiest to answer by making a graph? What kinds of questions would be easiest to answer by making a table? What kinds of questions would be easiest to answer by writing an equation?

   d. Is this relationship quadratic? Explain why or why not.

34. Suppose the circumference of a tree is x feet.

  a. What is the diameter of the tree in terms of x?

  b. What is the radius of the tree in terms of x?

  c. What is the area of a cross section of the tree in terms of x? Is the relationship between the circumference and the area of the cross section linear, quadratic, exponential, or none of these?

  d. If the circumference of the tree is 10 feet, what are the diameter, the radius, and the area of a cross section?

35. A square has sides of length x centimeters.

  a. The square is enlarged by a scale factor of 2. What is the area of the enlarged square?

  b. How does the area of the original square compare with the area of the enlarged square?

36. A rectangle has dimensions x centimeters and $x + 1$ centimeters.

  a. The rectangle is enlarged by a scale factor of 2. What is the area of the enlarged rectangle?

  b. How does the area of the original rectangle compare with the area of the enlarged rectangle?

Investigation 2: Quadratic Expressions 37

Extensions

37. A square has sides of length *x* centimeters. A new rectangle is created by increasing one dimension by 5 centimeters and decreasing the other dimension by 4 centimeters.

   a. Make a sketch to show how the square is transformed into the new rectangle.

   b. Write an expression for the area of the original square and an expression for the area of the new rectangle.

   c. For what *x* values is the area of the new rectangle greater than the area of the square? For what *x* values is the area of the new rectangle less than the area of the square? For what *x* values are the areas equal? Explain how you found your answers.

38. A square has sides of length *x* centimeters. A new rectangle is created by increasing one dimension by 2 centimeters and decreasing the other dimension by 3 centimeters.

   a. Make a sketch to show how the square is transformed into the new rectangle.

   b. Write two expressions, one in factored form and one in expanded form, for the area of the new rectangle.

   c. Write an equation for the area, *A*, of the new rectangle. Graph the equation, and describe the graph.

39. A square has sides of length *x* centimeters. A new rectangle is created by increasing one dimension by 2 centimeters and doubling the other dimension and then adding 2 centimeters.

   a. Make a sketch to show how the original square is transformed into the new rectangle.

   b. Write two expressions, one in factored form and one in expanded form, for the area of the new rectangle.

   c. Write an equation for the area, *A*, of the new rectangle. Graph the equation, and describe the graph.

In 40–43, the equation represents the area of a rectangle made by changing the dimensions of a square with sides of length x centimeters. Answer parts a and b.

40. $A = x^2 - 6x$ **41.** $A = x^2 - 9$

42. $A = x^2 + 8x + 16$ **43.** $A = x^2 + 10x + 16$

  a. Write an expression for the area in factored form.

  b. Sketch a graph of the equation, and describe the shape of the graph.

In 44–46, do parts a and b.

44. $y = x^2 - 2x - 8$ **45.** $y = x^2 - 5x$ **46.** $y = x^2 + 8x + 16$

  a. Match the equation with its graph.

  i.
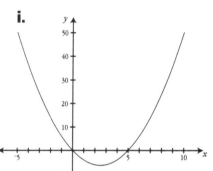

  ii.
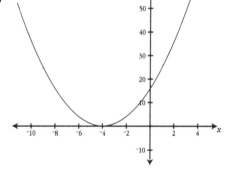

  iii.
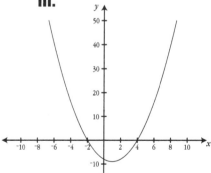

  b. Write each equation in factored form. Describe how you found the factored form.

Investigation 2: Quadratic Expressions 39

Mathematical Reflections

In this investigation, you explored quadratic equations representing the areas of rectangles formed by transforming a square. These questions will help you summarize what you have learned:

1. The quadratic expression $x(x + 7)$ is in factored form. How can you find an equivalent expression in expanded form? Write the expanded form for the expression, and sketch a rectangle to show that the two expressions are equivalent.

2. The quadratic expression $x^2 + 7x + 12$ is in expanded form. How can you find an equivalent expression in factored form? Write the factored form for the expression, and sketch a rectangle to show that the two expressions are equivalent.

3. In the graphs you made of the areas of the rectangles in this investigation, what information did the x-intercepts give you? How can you predict the x-intercepts for a quadratic function by looking at its equation?

4. How can you recognize a quadratic function from its equation?

5. Describe what you know about the shape of the graph of a quadratic function. Include the line of symmetry and any other important features you have observed.

Think about your answers to these questions, discuss your ideas with other students and your teacher, and then write a summary of your findings in your journal.

INVESTIGATION 3

Quadratic Patterns of Change

In earlier units, you studied linear relationships. In a linear relationship, the *y* value changes at a constant rate. This means that in a table of values, the difference in successive *y* values is constant. In this investigation, you will explore the patterns of change for quadratic relationships as you solve some interesting counting problems.

3.1 Counting Handshakes

After a sporting event, the opposing teams often line up and shake hands. To celebrate their victory, members of the winning team may congratulate each other with a round of high fives. In this problem, you will explore the total number of handshakes or high fives that take place in several situations. You will consider three cases.

Case 1
Two teams with the same number of players shake hands.

Case 2
Two teams with different numbers of players shake hands. For example, although 5 players from each basketball team participate at one time, one team may have a total of 8 players and the other may have a total of 10 players.

Case 3
Members of the same team exchange high fives.

Investigation 3: Quadratic Patterns of Change 41

Problem 3.1

A. Consider case 1, in which two teams have the same number of players. Each player on one team shakes hands with each player on the other team.

  1. How many handshakes will take place between two basketball teams with 10 players each?

  2. How many handshakes will take place between two soccer teams with 15 players each?

  3. Write an equation for the number of handshakes, h, between two teams with n players each.

B. Consider a restricted form of case 2, in which the numbers of players on the two teams differ by 1. Each player on one team shakes hands with each player on the other team.

  1. How many handshakes will take place between a water polo team with 10 players and a water polo team with 9 players?

  2. How many handshakes will take place between a field hockey team with 15 players and a field hockey team with 14 players?

  3. Write an equation for the number of handshakes, h, between a team with n players and a team with $n - 1$ players.

C. Consider case 3, in which each member of a team gives a high five to each teammate.

  1. How many high fives will take place among an academic quiz team with 4 members?

  2. How many high fives will take place among a golf team with 12 members?

  3. Write an equation for the number of high fives, h, that will take place among a team with n members.

Problem 3.1 Follow-Up

1. a. For cases 1, 2, and 3 in Problem 3.1, make a table showing the number of players on each team and the number of handshakes or high fives. Include data for teams with from 1 to 10 members. For case 2, consider only the case when the numbers of players on the two teams differ by 1.

  b. For each table, describe a pattern that would help you predict the number of handshakes or high fives for larger teams.

  c. Compare the patterns for the three tables. How are the patterns similar? How are they different?

2. **a.** Graph your three equations from Problem 3.1. Show *n* values from 1 to 10.
   b. How do your three graphs compare?
3. Look at the tables, graphs, and equations for the three cases. Are any of the relationships quadratic? Explain.
4. Tyler expressed the number of handshakes between a team with *n* members and a team with *n* − 1 members as $n(n-1)$. Asuko wrote $n^2 - n$ to represent the same situation.
   a. Describe how Tyler might have thought about the situation in order to formulate his expression. Describe how Asuko might have thought about the situation.
   b. Draw and label a rectangle whose area can be represented by these expressions.

3.2 Exploring Triangular Numbers

Study the pattern below.

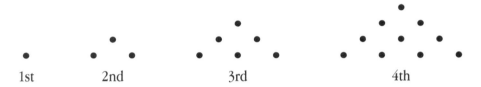

1st   2nd   3rd   4th

How many dots do you predict will be in the 5th figure? In the *n*th figure?

Problem 3.2

The numbers of dots in the figures above are called **triangular numbers.** The first triangular number is 1, the second triangular number is 3, the third is 6, the fourth is 10, and so on.

A. What two variables are important in this situation? Which is the independent variable, and which is the dependent variable?

B. Look for a pattern in the figures above. Use the pattern to help you make a table of the first ten triangular numbers.

C. Describe the pattern of change from one triangular number to the next.

D. How can you use this pattern of change to predict the 15th triangular number without making a drawing?

E. Write an equation that can be used to determine the *n*th triangular number.

F. Does your equation represent a quadratic relationship? Explain your answer.

Investigation 3: Quadratic Patterns of Change

Problem 3.2 Follow-Up

The pattern below represents the triangular numbers as collections of squares.

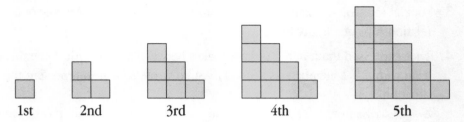

You can combine two copies of each figure to form a rectangle.

1. Give the dimensions and the area of the rectangle formed by combining two copies of the given figure.
   a. the 4th figure
   b. the 5th figure
   c. the 10th figure
   d. the nth figure

2. For each part of question 1, how does the number of squares in the rectangle compare with the number of squares in the original figure?

3. Use your answers from questions 1 and 2 to help you write an equation for calculating the nth triangular number. How does your equation compare with the equation you wrote in Problem 3.2?

4. What is the 18th triangular number?

5. Is 210 a triangular number? Explain your answer.

Applications • Connections • Extensions

As you work on these ACE questions, use your calculator whenever you need it.

Applications

1. In a school math league, each team has six student members and two coaches.

   a. At the start of a match, the coaches and student members of one team exchange handshakes with the coaches and student members of the other team. How many handshakes will be exchanged?

   b. At the end of the match, the members and coaches of the winning team exchange high fives to celebrate their victory. How many high fives will be exchanged?

2. The dot patterns below represent the first four *square numbers*, 1, 4, 9, and 16.

   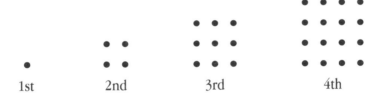

   a. What are the next two square numbers?

   b. Write an equation for calculating the *n*th square number.

   c. Make a table and a graph of the first ten square numbers. Describe the pattern of change from one square number to the next.

Investigation 3: Quadratic Patterns of Change 45

3. The dots below are arranged in rectangular patterns. The numbers of dots in the figures are called the *rectangular numbers*.

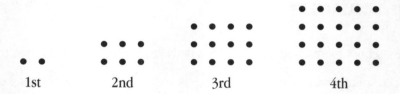

1st　　　2nd　　　3rd　　　4th

   a. What are the first four rectangular numbers?

   b. Find the next two rectangular numbers.

   c. Describe the pattern of change from one rectangular number to the next.

   d. Use the pattern of change to predict the 7th and 8th rectangular numbers.

   e. Write an equation for calculating the *n*th rectangular number.

4. In what ways are triangular numbers, square numbers, and rectangular numbers similar to the cases given in Problem 3.1?

In 5–8, tell whether the number is a triangular number, a square number, a rectangular number, or none of these, and explain how you know you are correct.

5. 110　　　**6.** 66　　　**7.** 121　　　**8.** 60

9. Graphs i–iv on the next page represent situations you have looked at in this unit. Study the graphs and then answer the following questions.

   a. Which graph might represent the number of high fives exchanged among a team with *n* players? Justify your choice.

   b. Which graph might represent areas of rectangles with a fixed perimeter? Justify your choice.

   c. Which graph might represent the area of a rectangle formed by increasing one dimension of a square by 2 centimeters and decreasing the other dimension by 3 centimeters? Justify your choice.

i.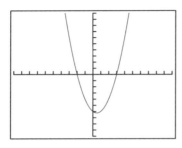
ii.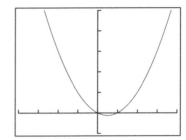
iii.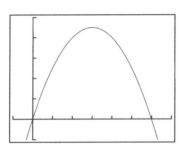
iv.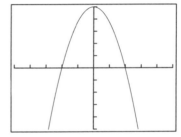

In 10–12, describe a situation that can be represented by the given equation, and tell what the variables P and n represent in that situation.

10. $P = n(n - 1)$ **11.** $P = n^2$ **12.** $P = n(n - 2)$

Connections

13. A *diagonal* is a line segment connecting any two nonadjacent vertices of a polygon. A quadrilateral has two diagonals.

    a. How many diagonals does a pentagon have? A hexagon? A heptagon? An octagon?

    b. How many diagonals does an *n*-sided polygon have?

Investigation 3: Quadratic Patterns of Change

14. These "trains" were formed by joining identical squares.

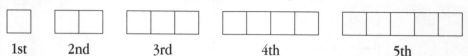

1st   2nd   3rd   4th   5th

a. How many rectangles can be found in each of the first five trains? For example, the shading below shows that there are 6 rectangles in the 3rd train. (Remember, a square is a special kind of rectangle.)

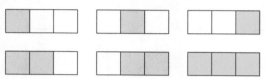

b. Make a table showing the number of rectangles in each of the first ten trains.

c. Describe how you could use the pattern of change in your table to find the number of rectangles in the 15th train.

d. Write an equation for calculating the number of rectangles in the nth train.

e. Use your equation to find the number of rectangles in the 15th train.

Extensions

15. You can quickly find the sum of the whole numbers from 1 to 10 by grouping them as shown below:

$$(1 + 10) + (2 + 9) + (3 + 8) + (4 + 7) + (5 + 6) = 11 + 11 + 11 + 11 + 11$$
$$= 5 \times 11$$
$$= 55$$

a. How could you use this idea to calculate $1 + 2 + 3 + \ldots + 99 + 100$?

b. How could you use this idea to calculate $1 + 2 + 3 + \ldots + n$ for any whole number n?

16. The dots below are arranged in patterns that represent the first three *star numbers*.

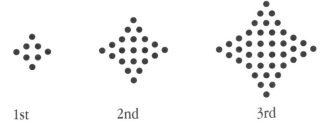

1st 2nd 3rd

a. What are the first three star numbers?

b. Use the dot patterns and your answers for part a to find the next three star numbers.

c. Write an equation you could use to calculate the *n*th star number.

17. a. Ten former classmates attend their class reunion. They all shake hands with each other. How many handshakes are exchanged? Explain your answer by drawing a picture or writing a convincing argument.

b. A little later, two more classmates arrive. If these two people shake hands with each other and the ten other classmates, how many new handshakes are exchanged? Explain your answer by drawing a picture or writing a convincing argument.

18. The dots below are arranged in hexagonal patterns. The numbers of dots in the patterns represent the *hexagonal numbers*.

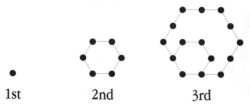

1st   2nd   3rd

a. What are the first three hexagonal numbers?

b. Use the dot patterns and your answers from part a to find the next two hexagonal numbers.

c. Which equation below could you use to calculate the *n*th hexagonal number?

$h = \frac{n}{2}(n + 1)$ $h = n(3n - 2)$ $h = n(2n - 1)$

19. If you look carefully, you can find 30 squares of various sizes in this 4-by-4 grid.

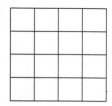

a. Sixteen of the 30 squares are the identical small squares that make up the grid. Find the other 14 squares. Draw pictures or give a description that would help someone else see all 30 squares.

b. How many squares can you find in an *n*-by-*n* grid? (Hint: Start with some simple cases and search for a pattern.)

Frogs, Fleas, and Painted Cubes

Mathematical Reflections

In this investigation, you counted handshakes and studied geometric patterns. You found that these situations could be represented by quadratic functions. These questions will help you summarize what you have learned:

1 In what ways are the relationships in the handshake problems similar to the relationships in the dot-pattern problems? In what ways are these relationships different?

2 In what ways are the quadratic functions in this investigation similar to the quadratic functions in Investigations 1 and 2? In what ways are they different?

3 a. Describe the patterns of change for the quadratic functions in this investigation.

b. In a table of values for a quadratic function, how can you use the pattern of change to predict the next entry?

Think about your answers to these questions, discuss your ideas with other students and your teacher, and then write a summary of your findings in your journal.

INVESTIGATION

What Is a Quadratic Function?

If you jump from a diving board or a bungee tower, gravity pulls you toward Earth. If you throw or kick a ball into the air, gravity brings it back down. For several hundred years, scientists have used mathematical models to describe and predict the effect of gravity on the position, velocity, and acceleration of falling and thrown objects.

In the first two problems of this investigation, you will look at how the height of a ball changes after it is thrown and how the distance from the ground of a person, a frog, and a flea change after they jump. You will look at tables, graphs, and equations for these situations and explore how changing the starting position affects the representations. In the last problem, you will review all you have learned so far about tables, graphs, and equations of quadratic functions, concentrating on the patterns of change in tables.

4.1 Tracking a Ball

No matter how hard you throw or kick a ball into the air, gravity always returns it to Earth. In this problem, you will explore how the height of a ball changes over time.

Problem 4.1

Suppose you throw a ball straight up into the air. This table describes how the height of the ball might change as it travels through the air.

Time (s)	Height (ft)
0.00	0
0.25	15
0.50	28
0.75	39
1.00	48
1.25	55
1.50	60
1.75	63
2.00	64
2.25	63
2.50	60
2.75	55
3.00	48
3.25	39
3.50	28
3.75	15
4.00	0

A. Describe how the height of the ball changes over this 4-second time period.

B. Without making the graph, describe what the graph of these data would look like. Include as many important features as you can.

C. Do you think these data represent a quadratic function? Explain why or why not.

Investigation 4: What Is a Quadratic Function?

Problem 4.1 Follow-Up

1. The height, h, of the ball after t seconds can be described by the equation
$$h = -16t^2 + 64t.$$

   a. Graph this equation on your calculator.
   b. Does the graph match your description from part B of Problem 4.1? Explain.
   c. Use the equation or the graph to figure out when the ball reaches a height of about 58 feet.
   d. Use the equation to find the height of the ball after 1.6 seconds.

2. In Problem 4.1, the initial height of the ball is 0 feet. This is not very realistic because it means that you would have to lie on the ground and release the ball without extending your arm. A more realistic equation for the height of the ball after t seconds is
$$h = -16t^2 + 64t + 6.$$

   a. Make a table and a graph for this quadratic function.
   b. The 6 in the equation above is called a constant term. A **constant term** is a term that does not contain a variable. How does the constant term affect the table and the graph? What information does the constant term in this equation give?
   c. Compare the graphs of the equations in questions 1 and 2. Discuss the similarities and differences in the following:
      i. the maximum height reached by the ball
      ii. the x-intercepts
      iii. the pattern of change in the height of the ball over time

54 Frogs, Fleas, and Painted Cubes

Did you know?

Aristotle, the ancient Greek philosopher and scientist, believed that heavier objects fall faster than lighter objects. This idea was widely accepted until the late 1500s when it was challenged by the great Italian scientist Galileo Galilei. Galileo is highly regarded for his work in astronomy, mathematics, and physics, but is perhaps best known for his devotion to the scientific method. The scientific method involves observing, developing a hypothesis based on observations, and then designing an experiment to test that hypothesis.

Galileo began to doubt Aristotle's idea when he observed a hailstorm. He noticed that large and small hailstones hit the ground at the same time. If Aristotle's idea were correct, this would happen only if the larger stones dropped from a greater altitude than the smaller stones, or if the smaller stones started falling before the larger stones. Galileo didn't think either of these explanations was probable.

To prove his theory that heavy and light objects fall at the same rate, Galileo performed an experiment. He reportedly climbed to the highest point he could find—the top of the Tower of Pisa—and dropped two objects simultaneously. Although they had different weights, the objects hit the ground at the same time.

4.2 Measuring Jumps

Many animals are known for their jumping abilities. Most frogs can jump several times their body length. Fleas are tiny, but they can easily leap onto a dog or a cat. Some humans have amazing jumping ability as well. Many professional basketball players have vertical leaps of more than 3 feet.

Suppose you filmed a frog, a flea, and a basketball player as they jumped straight up as high as possible. If you studied the films frame by frame, you would find that the time, t, in seconds and the height, H, in feet are related by equations similar to these:

$$\text{frog: } H = -16t^2 + 12t + 0.2$$
$$\text{flea: } H = -16t^2 + 8t$$
$$\text{basketball player: } H = -16t^2 + 16t + 6.5$$

Problem 4.2

A. Use your calculator to make tables and graphs of these three equations. Since a jump doesn't take much time, look at heights for time values between 0 seconds and 1 second. In your tables, use intervals of 0.1 second.

B. What is the maximum height reached by each jumper, and when is the maximum height reached?

C. How long does each jump last? Explain how you found your answer.

D. What do the constant terms 0.2 and 6.5 tell you about the frog and the basketball player?

■ Problem 4.2 Follow-Up

1. For each jumper, describe the pattern of change in the height over time, and explain how the pattern is reflected in the table and the graph.

2. Mr. Jain is a jewelry maker. He would like to increase his profit by raising the price of his jade earrings. However, he knows that if he raises the price too high, he won't sell as many earrings and his profit will decrease. Using records of past sales, a business consultant developed the equation $P = 50s - s^2$ to predict the monthly profit, P, for a given sales price, s.
   a. Make a table and a graph for this equation.
   b. What do the equation, the table, and the graph suggest about the relationship between price and profit?
   c. What price will bring the greatest profit?
   d. How does this equation compare with the equations in Problem 4.2?

56 Frogs, Fleas, and Painted Cubes

> **Did you know?**
>
> - The average flea weighs 0.000001 pound and is 2 to 3 millimeters long. It can pull 160,000 times its own weight and can jump 150 times its own length. This is equivalent to a human being pulling 24 million pounds and jumping 1000 feet!
> - There are 2375 known species and subspecies of fleas, and fleas are found on all land masses, including Antarctica.
> - Most fleas make their homes on bats, rats, squirrels, and mice.
> - The bubonic plague, which killed a quarter of Europe's population in the fourteenth century, was spread by rat fleas.
> - Flea circuses originated about 300 years ago and were popular in the United States a century ago.
>
> Source: *New York Times Magazine*, 22 October 1995, p. 18.

4.3 Putting It All Together

In this unit, you have used algebraic equations to model a variety of quadratic functions. You may have noticed some common characteristics of these equations. Below are some familiar equations.

$$\text{square numbers: } s = n^2$$
$$\text{rectangular numbers: } r = n(n + 1)$$
$$\text{area of a rectangle with a perimeter of 20: } A = (10 - l)l$$
$$\text{number of high fives: } h = \frac{n}{2}(n - 1)$$
$$\text{a flea jump: } H = {-16}t^2 + 8t$$

You have also observed patterns in the graphs of quadratic functions. These graphs, called *parabolas*, are symmetric U-shapes with a maximum point or a minimum point.

In the previous investigation, you began to explore patterns in the tables of quadratic functions. In this problem, you will look more closely at these patterns.

To understand a relationship between two variables, it helps to look at how the value of one variable changes each time the value of the other variable increases by a fixed amount. You have seen that for a linear relationship, the y value increases by a constant amount each time the x value increases by 1. For example, look at this table for the linear relationship $y = 3x + 1$. The calculations under "First differences" are the differences between consecutive y values.

$y = 3x + 1$

x	y
0	1
1	4
2	7
3	10
4	13
5	16

First differences
4 − 1 = 3
7 − 4 = 3
10 − 7 = 3
13 − 10 = 3
16 − 13 = 3

Since the y value increases by 3 each time the x value increases by 1, the first differences for $y = 3x + 1$ are all 3.

Now, let's look at a quadratic relationship. The simplest quadratic relationship is $y = x^2$, the rule for generating square numbers. In fact, the word *quadratic* is derived from the Latin word for "square." The table below shows that the first differences for $y = x^2$ are not constant.

$y = x^2$

x	y
0	0
1	1
2	4
3	9
4	16
5	25

First differences
1 − 0 = 1
4 − 1 = 3
9 − 4 = 5
16 − 9 = 7
25 − 16 = 9

However, look at what happens when we find the *second* differences.

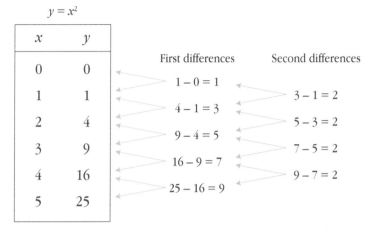

Do you think the tables for other quadratic functions will show a similar pattern?

Problem 4.3

A. Make a table for each quadratic equation below. Use integer values of x from -5 to 5. Add columns to your tables showing first and second differences.

$y = 2x(x + 3)$ $y = 3x - x^2$ $y = (x - 2)^2$ $y = x^2 + 5x + 6$

B. Consider the patterns of change in the y values and in the first and second differences for the four equations. In what ways are the patterns similar for the four equations? In what ways are they different?

Problem 4.3 Follow-Up

1. The patterns of differences for quadratic functions are not like the patterns of differences for the linear and exponential functions you studied in other units.
  a. Make a table of (x, y) values for each equation below. Include columns for the first and second differences.
  $y = x + 2$ $y = 2x$ $y = 2^x$ $y = x^2$
  b. For each table, look at the pattern of change in the y value as the x value increases by 1. How are the patterns similar in the four tables? How are they different?
  c. How are the patterns of change in the tables reflected in the equations?
2. You have seen that a parabola is a symmetric shape with either a maximum point or a minimum point. Describe the graph of each equation in Problem 4.3 and question 1 above. Be sure to consider maximum or minimum points, x-intercepts, and lines of symmetry. Use a calculator to check your descriptions.

Investigation 4: What Is a Quadratic Function?

Applications • Connections • Extensions

As you work on these ACE questions, use your calculator whenever you need it.

Applications

1. A basketball player takes a shot. The graph below shows the height of the ball, starting from when it leaves the player's hands.

   a. Estimate the height of the ball when it is released by the player.

   b. When does the ball reach its maximum height? What is the maximum height?

   c. How long does it take the ball to reach the basket, which is set at a height of 10 feet?

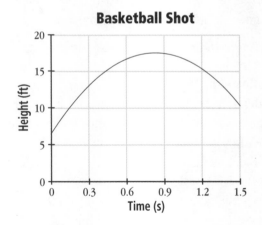

60 Frogs, Fleas, and Painted Cubes

2. The highest dive in the Olympic Games is from a 10-meter platform. The height in meters of a diver t seconds after leaving the platform can be estimated by the equation $h = 10 - 4.9t^2$.

   a. Make a table of the relationship between time and height.

   b. Sketch a graph of the relationship between time and height.

   c. When will the diver hit the water's surface? How can you find this answer by using your graph? How can you find this answer by using your table?

   d. When will the diver be 5 meters above the water?

   e. When is the diver falling fastest? How is this shown in the table and graph?

3. Kelsey jumps from a diving board, springing up into the air and then dropping feet-first toward the pool. The distance from her toes to the water's surface in feet t seconds after she leaves the board is $d = {}^-16t^2 + 18t + 10$.

   a. What is the maximum height of Kelsey's toes during this jump? When does the maximum height occur?

   b. When do Kelsey's toes hit the water?

   c. What does the constant term 10 in the equation tell you about Kelsey's jump?

In 4–8, tell whether the table represents a quadratic relationship. If the relationship is quadratic, tell whether it has a maximum value or a minimum value.

4.
x	-3	-2	-1	0	1	2	3	4	5
y	-4	1	4	5	4	1	-4	-11	-18

5.
x	0	1	2	3	4	5	6	7	8
y	2	3	6	11	18	27	38	51	66

Investigation 4: What Is a Quadratic Function?

6.

x	0	1	2	3	4	5	6	7	8
y	0	−4	−6	−6	−4	0	6	14	24

7.

x	−4	−3	−2	−1	0	1	2	3	4
y	5	4	3	2	1	2	3	4	5

8.

x	−4	−3	−2	−1	0	1	2	3	4
y	18	10	4	0	−2	−2	0	4	10

9. The distance from the ground of a ball in feet after t seconds can be described by the equation $h = {-}16t^2 + 48t + 8$.

   a. What is the maximum height attained by the ball? Explain how you could use a table and a graph of this relationship to find the answer.

   b. When does the ball hit the ground? Explain how you could use a table and a graph to find the answer.

   c. Describe the pattern of change in the height of the ball over time. Explain how this pattern would appear in a table and a graph.

   d. What does the constant term 8 mean in this context?

10. **a.** For each equation below, describe the pattern of change in the y value as the x value increases or decreases at a constant rate.

    $y = 2x^2$ $y = 3x^2$ $y = \frac{1}{2}x^2$ $y = {-}2x^2$

    b. What relationship do you see between the second difference and the number multiplied by the x^2 term in each equation in part a?

    c. Use what you discovered in parts a and b to predict the pattern of change for each of these equations.

    $y = 5x^2$ $y = {-}4x^2$ $y = \frac{1}{4}x^2$ $y = ax^2$

11. A signal flare is fired into the air from a boat. The height of the flare in feet t seconds after it is fired is $h = {}^-16t^2 + 160t$.

a. How high will the flare travel? When will it reach this maximum height?

b. When will the flare hit the water?

c. Explain how your answers to parts a and b could be found in a table and a graph of the equation.

In 12–15, do parts a–d.

12. $y = 9 - x^2$ **13.** $y = 2x^2 - 4x$

14. $y = 6x - x^2$ **15.** $y = x^2 + 6x + 8$

a. Sketch a graph of the equation.

b. Find the coordinates of the points where the graph crosses the x- and y-axes, and label these points on your graph.

c. Draw and label the line of symmetry.

d. Find the coordinates of the maximum or minimum point, and label this point on your graph.

16. A quadratic relationship between *x* and *y* is shown in the graph below.

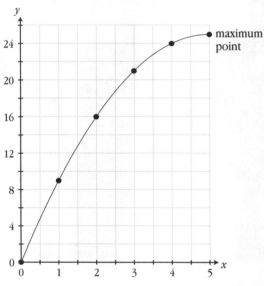

a. Make a table of (*x*, *y*) values for the six points shown.

b. Copy the graph, and extend it to show *x* values from 5 to 10. Explain how you know your graph is correct.

17. A quadratic relationship between *x* and *y* is shown in the graph below. Extend the *x*-axis to show values from −5 to 0. Explain how you know your graph is correct.

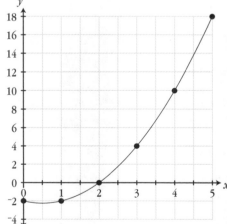

18. a. For each equation below, predict the shape of the graph. Give the maximum or minimum point, the x-intercepts, the y-intercepts, and the line of symmetry for each graph. Use a graphing calculator to check your predictions.

$y = x^2$ $\qquad$ $y = {-x^2}$ $\qquad$ $y = x^2 + 1$

$y = x^2 - 2$ $\qquad$ $y = x(4 - x)$ $\qquad$ $y = (x + 2)^2$

b. How can you tell from a quadratic equation whether the graph will have a maximum point or a minimum point?

c. How are the x- and y-intercepts of the graph of a quadratic function related to the equation?

Connections

19. A jeweler has tried to increase her profit by raising and lowering the price of her necklaces. Using past sales data, she has generated two equations relating income, i, to selling price, p:

$i = (100 - p)p$ and $i = 100p - p^2$

a. Are the two equations equivalent? How do you know?

b. Show that $i = 100 - p^2$ is not equivalent to the two original equations.

c. It costs $350 to rent a booth at a local art fair. Assume that the jeweler's profit for the fair will be her income from necklaces minus the cost of the booth. Write an equation for profit, M, as a function of selling price, p.

d. What price would give the maximum profit? What will the profit be for this price?

e. For what prices will there be a profit rather than a loss?

Investigation 4: What Is a Quadratic Function?

20. Use the table below to answer parts a and b.

x	−1	0	1	2	3	4	5
y	3	5	7	9	11		

  a. Describe the pattern in the table, and use the pattern to predict the missing y values.

  b. Tell whether the relationship between x and y is linear, exponential, quadratic, or none of these. Explain how you know.

21. Use the table below to answer parts a and b.

x	−1	0	1	2	3	4	5
y	−10	0	8	14	18		

  a. Describe the pattern in the table, and use the pattern to predict the missing y values.

  b. Tell whether the relationship between x and y is linear, exponential, quadratic, or none of these. Explain how you know.

22. Use the table below to answer parts a and b.

x	0	1	2	3	4	5	6
y	1	3	9	27	81		

  a. Describe the pattern in the table, and use the pattern to predict the missing y values.

  b. Tell whether the relationship between x and y is linear, exponential, quadratic, or none of these. Explain how you know.

23. Use the table below to answer parts a and b.

x	2	3	4	5	6	7	8
y	$\frac{1}{2}$	$\frac{1}{3}$	$\frac{1}{4}$	$\frac{1}{5}$	$\frac{1}{6}$		

a. Describe the pattern in the table, and use the pattern to predict the missing y values.

b. Tell whether the relationship between x and y is linear, exponential, quadratic, or none of these. Explain how you know.

24. Use the table below to answer parts a and b.

x	$^-1$	0	1	2	3	4	5
y	$^-5$	0	3	4	3		

a. Describe the pattern in the table, and use the pattern to predict the missing y values.

b. Tell whether the relationship between x and y is linear, exponential, quadratic, or none of these. Explain how you know.

Extensions

25. A puzzle involves a strip of seven squares, three pennies, and three nickels. To start, the pennies are placed on the left three squares and the nickels are placed on the right three squares.

The object of the puzzle is to switch the positions of the coins so that the nickels are on the left and the pennies are on the right. You can move a coin to an empty square by sliding it or by jumping it over one coin. You can move pennies only to the right and nickels only to the left.

You can make variations of this puzzle by using more or fewer coins and longer or shorter strips of squares. Each puzzle should have the same number of each type of coin and one empty square. You will explore these puzzles in parts a and b on the next page.

Investigation 4: What Is a Quadratic Function?

a. Make drawings that show each move (slide or jump) required to solve puzzles with 1, 2, and 3 coins of each type. How many moves does it take to solve each puzzle?

b. A puzzle with n nickels and n pennies can be solved with $n^2 + 2n$ moves. Use this quadratic expression to calculate the number of moves required to solve puzzles with 1, 2, 3, 4, 5, 6, 7, 8, 9, and 10 of each type of coin. Do the numbers of moves you found for 1, 2, and 3 coins of each type agree with the numbers you found in part a?

c. By calculating first and second differences in the data from part b, verify that the relationship between the number of moves and the number of each type of coin is quadratic.

In 26–28, use the following information: The soccer coach wants to take her team of 20 students to the state capital for a tournament. The travel agent says the trip will cost $125 per student, but the coach thinks this is too expensive. The travel agent suggests that the coach persuade other students to go with the team. For each extra student, the cost per student would be reduced by $1.

26. The trip actually costs the travel agent $75 per student. The remaining money is profit.

   a. What will the agent's profit be if only the 20 soccer team members go on the trip?

   b. What will the agent's profit be if 25 students go on the trip?

   c. What will the agent's profit be if 60 students go?

   d. What will the agent's profit be if 80 students go?

27. Let *n* represent the number of students who go on the trip. In a–d, write an equation for the relationship described. It may help to make a table like the one shown here and look for patterns in the columns.

Number of students	Price per student	Travel agent's income	Travel agent's expenses	Travel agent's profit
20	$125	20 × $125 = $2500	20 × $75 = $1500	$2500 − $1500 = $1000
21	124			

   a. the relationship between the price per student and *n*

   b. the relationship between the travel agent's income and *n*

   c. the relationship between the travel agent's expenses and *n*

   d. the relationship between the travel agent's profit and *n*

28. Make a table and a graph of the equation for the travel agent's profit. Study the pattern of change in the profit as the number of students increases from 25 to 75.

   a. What number of students gives the travel agent the maximum profit?

   b. What numbers of students will guarantee that the travel agent will earn a profit?

   c. What numbers of students will give the travel agent a profit of at least $1200?

Investigation 4: What Is a Quadratic Function?

Mathematical Reflections

In this investigation, you analyzed the relationship between height and time for several situations. You also looked for common features in the tables, graphs, and equations for quadratic relationships. These questions will help you summarize what you have learned:

1 Quadratic functions can be used to model many real-world situations. Describe three situations for which quadratic functions are appropriate models. For each situation, give examples of questions that quadratic representations help to answer.

2 What patterns of change occur in tables of (x, y) values for quadratic functions?

3 What patterns of change occur in graphs of quadratic functions?

4 How can you recognize a quadratic function from its equation?

Think about your answers to these questions, discuss your ideas with other students and your teacher, and then write a summary of your findings in your journal.

INVESTIGATION 5

Painted Cubes

Rubik's® Cube is a puzzle made of smaller cubes joined in the center by a special connector. When you first purchase a Rubik's Cube, each face is a different color: red, orange, green, white, blue, or yellow. The connector in the center allows you to twist the faces of the cube, mixing up the colors. Once you rearrange the colors, returning the puzzle to its original state can be very challenging.

Think about this!

The small cubes that make up Rubik's Cube are colored only on their exposed faces.

How many of the small cubes are colored on only one face? How many are colored on two faces? On three faces? On four faces? On five faces? On six faces?

5.1 Analyzing Cube Puzzles

As his midterm math project, Leon invented a puzzle. He made a large cube from 1000 centimeter cubes and then painted the faces of the large cube. When the paint dried, he separated the puzzle into the original centimeter cubes. The object of Leon's puzzle is to reassemble the cubes so that no unpainted faces are showing.

When Leon examined the centimeter cubes, he noticed that some were painted on only one face, some on two faces, and some on three faces. Many of the cubes weren't painted at all.

Problem 5.1

In this problem, you will investigate puzzles similar to Leon's.

A. 1. A cube with edges of length 2 centimeters is built from centimeter cubes. If you paint the faces of this cube and then break it into centimeter cubes, how many cubes will be painted on three faces? How many will be painted on two faces? On one face? How many will be unpainted?

2. Answer the questions from part 1 for cubes with edges of lengths 3, 4, 5, and 6 centimeters.

  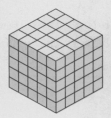 

Organize your data into a table like the one below.

Edge length of large cube	Number of cm cubes	Number of centimeter cubes painted on			
		3 faces	2 faces	1 face	0 faces
2					
3					
4					
5					
6					

B. Study the patterns in the table. Look for a relationship between the edge length of the large cube and the number of centimeter cubes. Tell whether the pattern of change is linear, quadratic, exponential, or none of these.

C. Look for relationships between the edge length of the large cube and the number of centimeter cubes painted on three faces, two faces, one face, and zero faces. Describe each relationship you find, and tell whether the pattern of change is linear, quadratic, exponential, or none of these.

■ **Problem 5.1 Follow-Up**

1. a. Use the patterns you discovered in the problem to extend your table to include data for cubes with edges of lengths 7, 8, and 9 centimeters.

   b. Make a graph of the data for each of the five relationships in the table.

2. What strategy would you use to reassemble the centimeter cubes to form the large cube so that no unpainted faces are showing?

5.2 Exploring Painted-Cube Patterns

In Problem 5.1, you discovered interesting relationships between the edge length of the large cube and the number of centimeter cubes painted on zero, one, two, or three faces. In this problem, you will write equations to generalize these relationships.

> **Problem 5.2**
>
> A large cube with edges of length n centimeters is built from centimeter cubes. The faces of the large cube are painted.
>
> **A.** Write an equation for the number of centimeter cubes in the large cube.
>
> **B.** Write an equation for the number of centimeter cubes painted on
>
> **1.** three faces **2.** two faces
>
> **3.** one face **4.** no faces

■ **Problem 5.2 Follow-Up**

1. a. Graph each of the five equations you wrote in Problem 5.2 on your calculator. Use the window settings shown. Make one graph at a time, and copy each onto your paper.

```
WINDOW
 XMIN=-10
 XMAX=10
 XSCL=1
 YMIN=-220
 YMAX=220
 YSCL=20
```

   b. How do the graphs you made in part a compare to those you made in Problem 5.1 Follow-Up?

Investigation 5: Painted Cubes

2. a. What is the relationship between the surface area, S, and the edge length, n, of the large cube? Write an equation for this relationship.

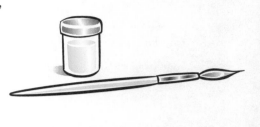

   b. For a cube with edge length n, use the equations you wrote in parts 1–3 of part B to help you write an equation for the relationship between the total number of painted faces, T, and n.

   c. What is the relationship between the surface area, S, and the total number of painted faces, T? Write an equation for this relationship.

3. a. Copy and complete the tables below.

x	x
1	1
2	
3	
4	
5	

x	x^2
1	1
2	
3	
4	
5	

x	x^3
1	1
2	
3	
4	
5	

   b. For each table, describe the pattern of change in the second column.

   c. Compare the patterns in these tables to the patterns in your table for the painted cubes. Match each table above with a column of the painted-cube table that shows a similar pattern of change, and explain how the two patterns are similar.

4. Compare the functions below with the functions represented by the tables in question 3. Match each equation with the table that shows a similar pattern of change, and explain how the two patterns are similar.

$$y = 2x \quad y = x^3 - 2 \quad y = (x - 1)^2$$

Applications • Connections • Extensions

As you work on these ACE questions, use your calculator whenever you need it.

Applications

1. A large cube with edges of length 12 centimeters is built from centimeter cubes. The faces of the large cube are painted.

   a. How many of the centimeter cubes are painted on three faces?

   b. How many of the centimeter cubes are painted on two faces?

   c. How many of the centimeter cubes are painted on one face?

   d. How many of the centimeter cubes are unpainted?

2. A large cube is built from centimeter cubes. The faces of the large cube are painted. Of the centimeter cubes, 1000 are unpainted.

   a. What is the length of an edge of the large cube?

   b. How many of the centimeter cubes are painted on one face?

   c. How many of the centimeter cubes are painted on two faces?

   d. How many of the centimeter cubes are painted on three faces?

3. A large cube is built from centimeter cubes. The faces of the large cube are painted. Of the centimeter cubes, 864 are painted on one face.

   a. What is the length of an edge of the large cube?

   b. How many of the centimeter cubes are unpainted?

   c. How many of the centimeter cubes are painted on two faces?

   d. How many of the centimeter cubes are painted on three faces?

4. A large cube is built from centimeter cubes. The faces of the large cube are painted. Of the centimeter cubes, 132 are painted on two faces.

   a. What is the length of an edge of the large cube?

   b. How many of the centimeter cubes are unpainted?

   c. How many of the centimeter cubes are painted on one face?

   d. How many of the centimeter cubes are painted on three faces?

5. A large cube is built from centimeter cubes. The faces of the large cube are painted. Of the centimeter cubes, 8 are painted on three faces.

   a. What is the length of an edge of the large cube?

   b. How many of the centimeter cubes are unpainted?

   c. How many of the centimeter cubes are painted on one face?

   d. How many of the centimeter cubes are painted on two faces?

6. A large cube with edges of length n centimeters is built from centimeter cubes. The faces of the large cube are painted.

   a. How many centimeter cubes were used to build the large cube?

   b. If 125 centimeter cubes were used, what is the value of n?

   c. If 343 of the centimeter cubes are unpainted, what is the value of n?

   d. If 120 of the centimeter cubes are painted on two faces, what is the value of n?

   e. If 486 of the centimeter cubes are painted on one face, what is the value of n?

7. The Tarheel Tile Company manufactures floor tiles with grid patterns based on this rule: corner squares are purple, edge squares that are not corners are gold, and inside squares are white. The illustration shows this pattern for a 4-by-4 grid.

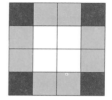

   a. If the same design rule is applied to a 5-by-5 grid, how many small squares will be purple? How many will be gold? How many will be white?

   b. If the same design rule is applied to a 10-by-10 grid, how many small squares will be purple? How many will be gold? How many will be white?

   c. If the same design rule is applied to a n-by-n grid, how many small squares will be purple? How many will be gold? How many will be white?

   d. Tell whether each relationship described in part c is linear, quadratic, exponential, or none of these.

8. The block pictured is made from centimeter cubes and then painted.

   a. How many of the centimeter cubes are painted on three faces?

   b. How many of the centimeter cubes are painted on two faces?

   c. How many of the centimeter cubes are painted on one face?

   d. How many of the centimeter cubes are unpainted?

   e. How many centimeter cubes does the block contain?

Investigation 5: Painted Cubes

9. A cube has edges of length x centimeters.

  a. Write an equation for the volume, V, of the cube in terms of x.

  b. If the length of each edge were doubled, how would the volume change?

  c. If the length of each edge were tripled, how would the volume change?

10. a. Copy and complete the table below.

Cube Data

Edge length (cm)	Surface area (cm²)	Volume (cm³)
0		
1		
2		
3		
4		
5		
6		
7		
8		
9		
10		
11		
12		

  b. Sketch a graph of the relationship between edge length and surface area.

  c. Sketch a graph of the relationship between edge length and volume.

  d. Does the relationship between edge length and surface area appear to be linear, quadratic, exponential, or none of these? Explain your answer.

  e. Does the relationship between edge length and volume appear to be linear, quadratic, exponential, or none of these? Explain your answer.

Connections

11. Some egg producers send eggs to supermarkets in 12-by-12-by-12 blocks.

   a. How many eggs are in one layer of a block?

   b. What is the total number of eggs in a block?

   c. What is the total number of eggs in a truckload 4 blocks wide, 4 blocks high, and 10 blocks long?

   d. Write an expression for the total number of eggs in a truckload n blocks long, m blocks wide, and 4 blocks high.

12. A square has sides of length x meters.

   a. Write equations for the area, A, and perimeter, P, of the square in terms of x.

   b. If the length of each side were doubled, how would the area change? How would the perimeter change?

   c. If the length of each side were tripled, how would the area change? How would the perimeter change?

   d. If the area of a square is 36 square meters, what is its perimeter?

13. a. Make a table showing the side length, perimeter, and area for squares with integer side lengths from 0 centimeters to 12 centimeters.

   b. Make a graph of the relationship between side length and perimeter for squares with side lengths from 0 centimeters to 12 centimeters.

   c. Make a graph of the relationship between side length and area for squares with side lengths from 0 centimeters to 12 centimeters.

   d. Does the relationship between side length and perimeter appear to be linear, quadratic, exponential, or none of these? Explain your answer.

   e. Does the relationship between side length and area appear to be linear, quadratic, exponential, or none of these? Explain your answer.

Investigation 5: Painted Cubes

In 14–17, do parts a–c.

14.

x	y
0	25
1	50
2	100
3	200
4	400
5	

15.

x	y
−3	3
−2	6
−1	9
0	12
1	15
2	

16.

x	y
2	6
3	12
4	20
5	30
6	42
7	

17.

x	y
−2	21
−1	24
0	25
1	24
2	21
3	

a. Describe the pattern of change in the table, and use the pattern to predict the missing entry.

b. Match the table with one of these equations.

$y = x^2 - 12$ $\qquad$ $y = 3x$ $\qquad$ $y = x(x + 1)$

$y = 25 - x^2$ $\qquad$ $y = (x)(x)(x)$ $\qquad$ $y = 3(x + 4)$

$y = 25(2^x)$

c. Does the table represent a quadratic relationship? How did you decide?

d. Does the relationship represented in the table have a maximum value?

e. Does the relationship have a minimum value?

In 18–21, tell whether the graph represents a quadratic relationship, and explain your reasoning. For those graphs that represent quadratic relationships, use a calculator to find an equation that matches the graph reasonably well. Record your equation and the window settings you used.

18.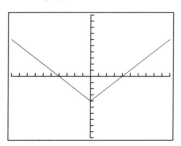

19.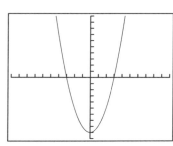

20.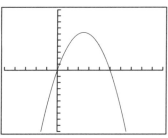

21.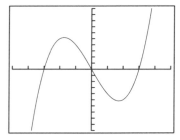

Extensions

22. Make a table of values and a graph for each function below. Include x values from -10 to 10. Describe the patterns you observe.

$y = x$ $y = x^2$ $y = x^3$

$y = x^4$ $y = x^5$ $y = x^6$

Investigation 5: Painted Cubes

23. Do parts a and b for each equation below.

$y = x + 1$ $\qquad\qquad y = (x + 1)(x + 2)$

$y = (x + 1)(x + 2)(x + 3)$ $\qquad y = (x + 1)(x + 2)(x + 3)(x + 4)$

a. Make a graph of the equation. Describe the shape of the graph, including important features.

b. Make a table of (x, y) values for the equation. Describe the pattern of change in the table.

24. A shell of white centimeter cubes is built around a large purple cube.

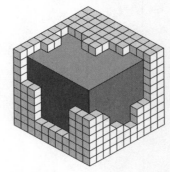

a. How many white cubes are needed if the purple cube has edges of length 2 centimeters? Explain.

b. How many white cubes are needed if the purple cube has edges of length 7 centimeters? Explain.

c. Write an equation for calculating the number of white cubes, W, needed to cover a purple cube with edges of length n centimeters.

d. Make a table and a graph of your equation for n values from 0 to 12.

e. If the white shell contains 218 cubes, what is the edge length of the purple cube?

f. If the white shell contains 602 cubes, what is the edge length of the purple cube?

Frogs, Fleas, and Painted Cubes

25. Centimeter cubes are stacked in the following pattern.

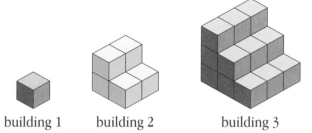

building 1   building 2   building 3

a. How many cubes will be in building 4? How many cubes will be in building 5?

b. Write an equation for the relationship between the building number and the number of cubes in the building.

c. Is the relationship between the building number and the number of cubes quadratic? Explain.

Mathematical Reflections

In this investigation, you explored the relationships involved in cube puzzles. These questions will help you summarize what you have learned:

1 a. Which of the relationships involved in the cube puzzles are linear?

   b. What common patterns occur in tables, graphs, and equations for linear relationships?

2 a. Which of the relationships involved in the cube puzzles are quadratic?

   b. What common patterns occur in tables, graphs, and equations for quadratic relationships?

3 a. Which of the relationships involved in the cube puzzles are neither linear nor quadratic?

   b. What are some patterns you observed in tables, graphs, and equations for these relationships?

Think about your answers to these questions, discuss your ideas with other students and your teacher, and then write a summary of your findings in your journal.

Glossary

constant term A number in an algebraic expression that is not multiplied by a variable. In the expanded form of a quadratic expression, the constant term is the number that is added. The constant term in the expression $-16t^2 + 64t + 7$ is 7, and the constant term in the expression $x^2 - 4$ is -4.

expanded form The form of an expression composed of sums or differences of terms rather than products of factors. The expressions $x^2 + 7x + 12$ and $x^2 + 2x$ are in expanded form.

factored form The form of an expression composed of products of factors rather than sums or differences of terms. The expressions $(x + 3)(x + 4)$ and $x(x - 2)$ are in factored form.

function A relationship between two variables in which the value of one variable depends on the value of the other variable. The relationship between length and area for rectangles with a fixed perimeter can be thought of as a function, where the area of the rectangle depends on, or is a *function of,* the length. If the variable y is a function of the variable x, then there is exactly one y value for every x value.

like terms Terms with the same variable raised to the same exponent. In the expression $4x^2 + 3x - 2x^2 - 2x + 1$, $3x$ and $-2x$ are like terms and $4x^2$ and $-2x^2$ are like terms.

line of symmetry A line that divides a graph or drawing into two halves that are mirror images of each other.

linear term A part of an algebraic expression in which the variable is raised to the first power, especially in the expanded form of an expression. In the expression $4x^2 + 3x - 2x^2 - 2x + 1$, $3x$ and $-2x$ are linear terms.

maximum value The greatest y value of a function. If y is the height of a thrown object, then the maximum value of the height, or simply the maximum height, is the greatest height the object reaches. If you throw a ball into the air, its height increases until it reaches the maximum height, and then its height decreases as it falls back to the ground. If y is the area of a rectangle with a fixed perimeter, then the maximum value of the area, or simply the maximum area, is the greatest area possible for a rectangle with that perimeter. In this unit, you found that the maximum area for a rectangle with a perimeter of 20 meters is 25 square meters.

minimum value The smallest y value of a function. If y is the cost of an item, then the minimum value of the cost, or simply the minimum cost, is the least cost possible for that item.

parabola The graph of a quadratic function. A parabola has a line of symmetry that passes through the maximum point if the graph opens downward or through the minimum point if the graph opens upward.

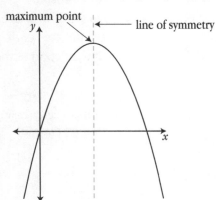
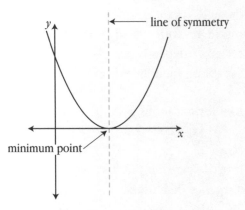

quadratic expression In expanded form, an expression with quadratic terms and possibly constant or linear terms as well. In factored form, a quadratic expression has two linear factors. The quadratic expression $x^2 - 2x$ is in expanded form. The equivalent expression in factored form is $x(x - 2)$. Quadratic expressions are often used to describe situations in two dimensions, such as area.

quadratic term A part of an algebraic expression in which the variable is raised to the second power, especially in the expanded form of an expression. In the expression $4x^2 + 3x - 2x^2 - 2x + 1$, $4x^2$ and $-2x^2$ are quadratic terms.

term An expression with numbers and/or variables multiplied together. In the expression $3x^2 - 2x + 10$, $3x^2$, $-2x$, and 10 are terms.

triangular number A quantity that can be arranged in a triangular pattern. The first four triangular numbers are 1, 3, 6, and 10.

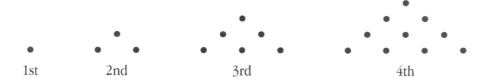

1st   2nd   3rd   4th

Index

Aristotle, 55

Constant term, 54, 85
Cube patterns, 71–74
    ACE, 75–83

Dependent variable, 18
Diagonal, of a polygon, 47

Equation, 21
    quadratic, 24
    for a quadratic relationship, 10–11
Equivalent expression, 22
Expanded form, 23, 85
Expression, 21
    equivalent, 22

Factored form, 23, 85
Function, 7, 85

Galilei, Galileo, 55
Glossary, 85–87
Graph, of a quadratic relationship, 7–9
Graphing calculator, comparing graphs of quadratic equations, 29–30

Hexagonal number, 50

Independent variable, 18

Investigation
    Introduction to Quadratic Relationships 5–18
    Painted Cubes, 71–84
    Quadratic Expressions, 19–40
    Quadratic Patterns of Change, 41–51
    What Is a Quadratic Function, 52–70

Journal, 18, 40, 51, 70, 84

Like terms, 27
    combining, 27
Linear term, 85
Line of symmetry, 28, 85

Mathematical Highlights, 4
Mathematical Reflections, 18, 40, 51, 70, 84
Maximum value, 28, 86
Minimum value, 28, 86

Parabola, 4, 7, 11, 86
    shape in relation to quadratic function, 28–30
Patterns of change
    ACE, 45–50
    for quadratic relationships, 41–44
Patterns of differences
    for linear functions, 57–58
    for quadratic functions, 57–59

Quadratic equation, 24
    ACE, 31–39
    parabola shape in relation to, 28–30
Quadratic expression, 4, 19–30, 86
    ACE, 31–39
Quadratic function, 52–59
    ACE, 31–39, 45–50, 60–69
    parabola shape in relation to, 28–30
    patterns of change for, 41–44
    patterns of differences for, 57–59
    time and height relationships, 52–57
Quadratic relationship, 4, 7
    ACE, 12–17, 45–50, 60–69, 75–83
    cube patterns, 71–74
    equation for, 10–11
    graph of, 7–9
    length and area of rectangles with a fixed perimeter, 6–11
    patterns of change for, 41–44
    time and height, 52–57
Quadratic term, 87

Rectangle, maximum area for fixed perimeter, 6–11
Rectangular number, 46
Rubik's Cube, 71

Square number, 45
Symbolic expression, 21

Term, of an expression, 23, 87
Triangular number, 43–44, 87

x**-intercept,** 28

y**-intercept,** 28